Backyard Beekeeping in the Peach State

A No-Fluff Guide for Beginner Beekeepers

By Radiah Mallard

Dedication

To my Heavenly Father—YAHUAH, the Creator of Heaven and Earth —thank YOU for the wisdom in YOUR design and the beauty of YOUR creation. The bees, the garden, the soil, and every living thing declare YOUR glory. YOU have given us dominion not as owners, but as stewards, and I am humbled to be entrusted with even a small part of YOUR wondrous work. May this book reflect the reverence I have for what YOUR hands have made.

To my husband, Robert—my covering, my calm, my constant. Your love, strength, and unwavering support make it possible for me to walk out this calling. I see God's goodness in you daily.

To my children—each of you is a blessing and a testimony. I pray that through this journey, you see the value of hard work, faith, and walking in purpose together.

To my dear friends who stood by me when this vision was only a whisper—thank you for your prayers, encouragement, and belief in what could be.

To Ms. Mary at the community garden—your kindness and patience planted seeds in more ways than one. You welcomed us, taught us, and loved us through the learning.

And to Ronzelle Buckner—thank you for opening the gate, both literally and spiritually, to this journey of beekeeping. Your willingness to let us learn and grow has changed our lives and given birth to something lasting.

This book is for all who believe in nurturing what's been entrusted to them—with care, compassion, and purpose. May it bring honor to the Creator and hope to the keeper.

Published by: DBORAH CONCEPTS LLC

EBOOK ISBN: 978-1-968508-04-3

PAPERBACK ISBN: 978-1-968508-03-6

30 N. Gould St

Ste. R

Sheridan, Wyoming 82801

United States

Website: www.dborahconcepts.co

Foundation Honey Co

Donate

Foreword

When I first stepped into a bee suit, I wasn't looking for healing—I was looking for quiet.

As a veteran, wife, and mother of four, life had demanded much of me. I knew what it meant to lead, to serve, to push through the hard seasons. But I also knew what it meant to feel the ache of invisible wounds—wounds that can linger long after the battle ends.

What I didn't expect was that beekeeping would give me back pieces of myself I didn't even know were missing.

This book is more than a guide to bees—it's an invitation. An invitation to see the hive as a place of restoration, reflection, and renewal. Especially for veterans, beekeeping offers something rare and sacred: peace. In the steady hum of the colony, there's a rhythm that calms the mind and brings the heart home. For those of us who've worn the uniform, we understand the weight of service and the struggle of transition. Coming back to civilian life can feel disorienting. You go from a life of structure and purpose to one that often feels scattered, misunderstood, or lonely. But the hive? The hive has structure. It has order. It has community. And it reminds you—without words—that you still have a mission.

Beekeeping grounds you in the present moment. You can't rush a hive. You can't show up angry or anxious. The bees demand your attention, your respect, your gentleness. In return, they give you something immeasurable: the space to be still. For many veterans, that alone can be the beginning of healing. But the beauty of beekeeping doesn't stop with the keeper. It spills over into the family. Into the neighborhood. Into the community.

I've seen what happens when a veteran invites their child to inspect a frame of honeycomb or teaches a neighbor how pollination works. I've seen spouses light up as their partner finds joy and confidence in something new. I've watched once- quiet gardens turn into gathering places—full of laughter, learning, and life because...

Beekeeping creates connection.

In historically underserved communities like ours in Columbus, Georgia, bees aren't just pollinators—they're powerful agents of change. They help turn food deserts into fresh harvests. They help veterans become educators, leaders, and advocates for sustainability. And they remind us that nature doesn't need to be tamed—it needs to be understood and respected.

This is personal for me. As a mother to a disabled son and a wife to a fellow veteran, I know the challenges of balancing care, service, and self. Beekeeping became a shared language in our family. A way to slow down, work together, and reconnect with what really matters. It became therapy with wings.

Scientifically, the sound of bees has been shown to calm the nervous system. The repetitive, focused tasks of beekeeping can ease symptoms of PTSD, depression, and anxiety. But you don't need a study to tell you what you'll feel the first time you open a hive: the quiet power of something greater than yourself at work.

That's why I wrote this book—not just to share the how, but the why. Why beekeeping matters. Why it heals. Why it belongs in our homes, our communities, and especially in the lives of our veterans.

If you're picking up this book, I believe you're ready to explore something deeper. Maybe you're a veteran searching for purpose. Maybe you're a family looking to reconnect. Maybe you're simply curious about what bees can teach us.

Whatever brought you here, know this: there is room for you in the hive.

"Sometimes, the smallest creatures lead us back to the biggest parts of ourselves."-Radiah Mallard, US Army, Retired

Table of Contents

Introduction

Have you ever taken a moment to watch a bee buzzing from flower to flower? If not, I encourage you to pause next time you see one. You'll notice how focused and deliberate each tiny bee is. Bees aren't just busy; they're crucial. They are nature's unsung heroes, quietly working behind the scenes, making sure we have food on our tables.

You might wonder, "Why should I care about bees?" It's simple: no bees, no food. Bees pollinate around one-third of everything we eat—from juicy peaches and crisp apples to hearty almonds and refreshing watermelons. In fact, Georgia's peaches wouldn't exist without bees doing their job. That's how important these little insects are. But right now, bees are in trouble, and they need our help.

When I first started beekeeping here in the Peach State, I was curious but overwhelmed. There was too much complicated information that felt like science class all over again. I didn't need more confusion; I needed clear instructions and confidence. I wanted a simple, no-fuss way to care for bees, harvest delicious honey, and make a positive difference.

So, I decided to cut through all the confusing jargon and share exactly what you need—straightforward and easy-to-understand. My mission with this guide is simple: introduce you to the rewarding world of backyard beekeeping without overwhelming you. By the time you finish this short, practical book, you'll feel prepared, confident, and excited to welcome bees into your backyard.

I know from personal experience how rewarding this journey is. My own beekeeping adventure began with a small curiosity and quickly turned into a deep passion for helping pollinators thrive. Along the way, I discovered that caring for bees isn't just about honey; it's about creating healthier communities, safer environments, and sustainable food sources.

This book is designed especially for Georgians who care deeply about their environment, food systems, and local communities. Whether you live in the heart of Columbus or have a rural plot of land, bees need you—and you can help.

First, let's clear up a common misunderstanding: Beekeeping isn't scary. Bees aren't aggressive by nature—they're gentle creatures who prefer flowers to people. Learning how to approach your bees with care and respect makes all the difference. It's surprisingly calming and therapeutic. Imagine sitting quietly in your backyard, sipping your tea, listening to the calming hum of your own bees doing their daily work.

In this guide, I'll introduce you to easy-to-manage hive styles, starting with the horizontal hive. I love this hive because it requires minimal lifting and offers simple management—perfect for beginners. I'll also briefly introduce you to two other hive types you might research later: the common Langstroth hive, and the natural, low-maintenance Top-Bar hive. Each hive style has its benefits, and you'll easily discover which suits your lifestyle best.

We'll also cover basic tools every new beekeeper needs. You don't have to spend a fortune or fill a shed with unnecessary equipment. A simple set of essential tools can get you started safely and affordably. I'll guide you through sourcing your first bees from local, reputable suppliers, ensuring your bees are healthy and ready for life in your Georgia backyard.

As you embark on this exciting journey, you'll also become an advocate for chemical-free beekeeping. Our bees, our families, and our communities deserve honey that's free from chemicals and pesticides. Keeping bees naturally isn't just possible—it's easier and better for everyone involved. You'll learn simple, effective ways to keep your bees happy and healthy without toxic substances. I'll guide you through gentle, natural methods to manage common hive problems, making sure your bees and your honey remain pure and healthy.

We'll also talk about creating a bee-friendly space in your backyard, with easy plant choices that bees adore. Whether you have a tiny yard or a large garden, even a few simple flowers make a big difference. Bees need safe, welcoming places to thrive, and you can provide that space—no matter where you live.

As you read through each chapter, you'll gain practical knowledge to confidently harvest honey that tastes amazing and is completely organic. There's nothing quite like sharing fresh honey from your backyard hive with family, friends, and neighbors. You'll discover how simple and rewarding harvesting honey can be—another sweet benefit of becoming a backyard beekeeper.

Finally, we'll discuss simple yet powerful steps you can take to advocate for bees beyond your backyard. Whether promoting pollinator-friendly practices in your community, encouraging chemical-free landscaping in your neighborhood, or simply sharing your honey and your experiences, you'll become part of something much bigger—a movement to protect bees and our environment, one backyard at a time.

My greatest hope is that by reading this book, you'll feel excited, confident, and ready to welcome bees into your life. You don't need to be an expert or have special skills. All you need is curiosity, care, and commitment. Remember, this journey doesn't stop when you close this guide; it's just beginning.

Together, let's nurture healthier communities, cleaner environments, and thriving bees. The Peach State needs backyard beekeepers like you, and you've already taken the first step. Welcome to the wonderful, buzzing world of beekeeping—Georgia style!

Chapter 1
Understanding Bees and Their Impact

Bees: Nature's Tiny Helpers

Did you know that every third bite of food you eat depends on bees? These tiny buzzing creatures might seem small and simple, but their impact is huge. In Georgia, where peaches, blueberries, apples, and watermelons grow abundantly, bees are essential to keeping our tables full and colorful.

Bees are nature's unsung heroes. They quietly go about their business of pollinating flowers, fruits, and vegetables, making sure plants produce seeds and grow healthy food. Without bees, our diets would be much less tasty—and our farms less productive.

In short, bees keep our world buzzing and our gardens blooming.

Meet the Bee Family

In every hive, there's a buzzing community of bees that works as a team. Three main types of bees live together in each colony: the queen, the workers, and the drones. Let's meet these amazing insects:

The Queen Bee Every hive has just one queen, and she's the star of the colony. Her main job is laying eggs—up to 2,000 eggs a day during busy seasons! The queen is larger than other bees and lives longer, usually about 2-3 years. She releases special smells, called pheromones, that keep the hive peaceful and organized. Without the queen, the hive cannot survive. As a new beekeeper, you'll quickly learn that a healthy queen means a healthy hive.

Worker Bees Worker bees are the hardworking heroes of the hive. These bees are all female, and their jobs change throughout their short lives of about 5-6 weeks. When they're young, they clean the hive, take care of baby bees, and tend to the queen. Later, worker bees leave the hive to gather pollen, nectar, and water. They're also the hive's defenders, protecting their home from threats. Worker bees are the heart of your hive, and watching their teamwork is inspiring.

Drone Bees Drones are male bees, and their only job is to mate with a queen. Drones don't gather pollen, don't build the hive, and don't have stingers. They live carefree lives—until mating season ends. At that point, drones usually leave the hive or are gently escorted out by worker bees. Although they have a simple life, drones are essential because they keep the bee population healthy by ensuring strong genetics.

How Bees Support Our Food Supply

Imagine summer without juicy peaches or sweet strawberries. Bees make these delicious treats possible. As bees fly from flower to flower, they transfer pollen, helping plants produce fruit and seeds. Farmers rely on bees to pollinate crops that feed both people and animals.

Here in Georgia, bees contribute millions of dollars to our state's agriculture every year. From orchards to backyard gardens, these tiny pollinators keep food growing. Even crops that don't directly depend on bees, like cotton and peanuts, benefit from healthier soil and balanced ecosystems bees help maintain.

Bees and Georgia's Agriculture

Georgia is known for its diverse farms—peaches, blueberries, squash, cucumbers, melons, apples, pecans, and more. Each of these crops thrives with bees around. In fact, many Georgia farmers partner with local beekeepers, bringing hives directly to their fields.

By becoming a backyard beekeeper, you're helping Georgia's entire farming community. Even just one hive in your yard supports gardens and crops miles away. Your bees will happily pollinate flowers around your neighborhood, creating better harvests for everyone.

Exploring Further

Now you've met bees up close. The queen, workers, and drones each play special roles, ensuring that the colony survives and thrives. As a beginner beekeeper, you'll discover these fascinating creatures have personalities and habits of their own. You'll find yourself spending hours happily watching bees buzz in and out, amazed by their dedication and teamwork.

Your journey has just begun—let bees guide you into a new world of learning and curiosity. Remember, the more you know about bees, the more fascinating they become.

Let's keep exploring, and keep Georgia buzzing!

Chapter 2
Getting Started–Keeping it Simple

You've learned how important bees are, and maybe you're already imagining your own backyard buzzing with life. Getting started with beekeeping doesn't have to be complicated or expensive. In fact, it's easier than you might think. Let's keep it simple and dive right in!

Basic Beekeeping Terms (No confusing jargon!)

When you first look into beekeeping, some terms might seem confusing. Don't worry—I've got your back! Here are some easy-to-understand terms you'll hear frequently:

- **Apiary**: This is just a fancy word for the place where you keep your beehives. Your backyard is your apiary!
- **Colony**: A bee colony is like a big family, with a queen, workers, and drones all living together in the hive.
- **Brood**: Bee babies! Brood refers to the eggs, larvae, and pupae growing inside the hive.
- **Frame**: This is a removable piece that bees build their honeycomb on. Frames make it easy for you to inspect and harvest honey.
- **Comb**: Beeswax structures built by bees, holding honey, pollen, and baby bees.

Now, you're ready to talk bees with confidence—no fancy words required.

Essential Tools for New Beekeepers

Starting beekeeping doesn't require a shed full of equipment. Here are the basic tools you'll need to get started without overwhelming your wallet:

- **Hive**: Your bees' home. You'll learn more about types of hives in the next chapter.
- **Bee Suit or Jacket**: Keeps you protected when you're visiting your bees. If a full suit seems intimidating, a bee jacket and veil are enough to keep you safe and comfortable.
- **Gloves**: Protect your hands. Many beginners prefer thin gloves to feel comfortable working with bees.
- **Smoker**: A small, handheld device that produces smoke, calming the bees when you inspect your hive.
- **Hive Tool**: A simple, sturdy tool used to gently pry open your hive and remove frames easily.
- **Bee Brush**: Helps you gently brush bees off frames when checking or harvesting honey.

You might already have some of these items at home, and others are easy and affordable to buy online or at your local supply store.

Choosing Your First Bees (Local Sources

Getting your bees locally is always best. Local bees are used to Georgia's climate and are more likely to thrive. Here are three easy ways to get bees for your hive:

Package s A package is a small wooden box filled with worker bees and a queen. You can buy these packages from trusted local bee suppliers. Packages are easy to install, and instructions usually come right with the bees.

Nucleu2s. Colonies (Nucs) A nuc (pronounced "nuke") is a tiny hive already set up with frames, brood, workers, and a queen. It's like buying a small bee family that's already settled. Nucs are perfect for beginners who want a quick start.

Swarms (Free Bees!) Sometimes bees naturally leave overcrowded hives and look for new homes—this is called swarming. Local beekeeping groups can help you find free swarms. Catching swarms is exciting, but might require help from experienced beekeepers the first time you try.

Whichever way you choose, getting bees is an exciting moment and a big step toward becoming a real beekeeper.

Connecting with Local Beekeeping Groups

Beekeeping isn't something you have to figure out alone. One of the best resources available to new beekeepers in Georgia is local beekeeping groups. They're friendly, supportive, and eager to share their wisdom. Connecting with experienced beekeepers gives you a safety net and makes beekeeping even more fun.

- **Georgia Beekeepers Association**: They offer classes, workshops, and mentorship programs.
- **Local Clubs**: Search online for "beekeeping groups near me" or ask around at local farmers' markets.

Joining a group gives you the confidence you need and introduces you to a community that genuinely cares about bees and each other.

Keeping it Fun and Simple

Remember, the goal of backyard beekeeping is enjoyment, learning, and helping your community—not stress! You don't have to know everything right away. Beekeeping is something you'll get better at over time. Mistakes happen, and that's okay—your bees are incredibly forgiving teachers.

As you start your beekeeping journey, keep it fun, keep it simple, and never be afraid to ask questions or seek help. Every great beekeeper began exactly where you are right now.

You're doing something amazing: protecting bees, supporting Georgia's agriculture, and providing your family with delicious, natural honey. Not bad for a beginner, right?

Let's keep going. Your bees—and your exciting new adventure— are waiting!

Chapter 3
Choosing Your Hive

Why the Hive Matters

When my husband and I first started beekeeping, we were excited but quickly learned a valuable lesson—bees and honey are heavier than they look! As disabled veterans, we needed to find a beekeeping method that wouldn't strain our bodies. I served by jumping out of airplanes, and my husband was injured in combat, so lifting heavy boxes from traditional stackable hives quickly became a challenge.

After that first exhausting year with traditional Langstroth hives, we knew there had to be a better way. That's when our search led us to the Horizontal (Layens) hive. This discovery was a game-changer. Though a bit more expensive upfront, we saw it as a smart investment in our health and longevity as beekeepers. It quickly became our favorite hive, and we've been hooked ever since.

Your hive matters—a lot. The right hive can mean the difference between joyful, stress-free beekeeping and frustration or injury. Let's explore the Horizontal hive first, then look at two other easy options so you can choose what fits your needs best.

Horizontal Hives: Simple and Effective (Our Favorite!)

A Horizontal hive, also known as the Layens hive, is exactly what it sounds like—a long hive that sits at waist level. Instead of stacking heavy boxes vertically like traditional hives, this hive grows horizontally, allowing easy access without lifting heavy loads.

Why We Love Horizontal Hives:

- **Less lifting:** You simply pull frames out from the side, never lifting more than one frame at a time. This is much gentler on your back and knees.
- **Easy inspections:** Quick and comfortable hive checks mean less stress for you and your bees.
- **Better bee health:** Because these hives mimic natural bee habitats, your bees experience less stress and often stay healthier.
- **Accessibility:** Ideal for anyone with mobility challenges, like us, making beekeeping fun rather than physically demanding.

Yes, Horizontal hives cost a bit more initially. But we quickly learned that investing in our health and comfort was worth every penny. It allowed us to stay active, joyful, and pain-free while doing what we loved—caring for bees.

Two More Hive Options to Explore

While we've shared why Horizontal hives are perfect for our lifestyle, every beekeeper is different. Let's briefly explore two more popular hive types you might want to research further:

Langstroth Hive: Common and Practical

The Langstroth hive is probably what comes to mind first when you think of beekeeping. It's the most common hive in America and consists of stacked boxes with removable frames.

Pros:

- Easy to find, affordable, and widely used.
- Parts are standardized and easy to replace.
- Good for commercial honey production.

Cons (for us):

- Boxes full of honey and bees get heavy (up to 50 lbs!).
- Frequent lifting required for inspections.

- Challenging for those with back, joint, or strength limitations.

Even though it didn't fit our needs, Langstroth hives might still work great for you—especially if you're young, strong, or planning a smaller setup.

Top-Bar Hive: Natural and Low-Maintenance

The Top-Bar hive is another horizontal-style hive, shaped somewhat like a trough or wooden bathtub. Bees build comb naturally along bars at the top, which you lift out individually.

Pros:

- Easy inspections without heavy lifting.
- Affordable and easy to build yourself.
- Bees create natural comb, promoting bee health.

Cons:

- Slightly less honey production compared to Langstroth hives.
- Frames are delicate and need careful handling.

Top-Bar hives are a great middle ground between traditional Langstroth and Horizontal Layens hives, especially for hobbyists who want easy management without heavy lifting.

Choosing the Best Hive for You

When picking a hive, consider your physical abilities, goals, budget, and how much honey you want to harvest. For us, the Horizontal (Layens) hive checked all the boxes:

- No heavy lifting ✅
- Less physical stress ✅
- Easy hive management ✅
- Investment in long-term health ✅

Our journey started with lifting heavy Langstroth boxes, but now we enjoy relaxed afternoons tending our Horizontal hives without worrying about pain or injuries. It has truly made beekeeping accessible, fun, and sustainable for us as disabled veterans.

A Hive That Fits Your Life

Remember, the right hive makes beekeeping enjoyable, not stressful. Do your research, consider your health, and don't be afraid to try something new. Your hive is more than just a home for bees—it's part of your lifestyle and your joy.

If you're like us—facing physical limitations but still determined to keep bees—consider a Horizontal hive. The initial investment pays off every day in comfort and peace of mind.

Whichever hive you choose, the goal is always the same: keep bees healthy, enjoy the journey, and support your community and environment.

Your perfect hive is out there. Find it, and you'll love every minute you spend keeping bees buzzing.

Now that you know about hives, let's talk about creating bee-friendly backyard that will keep your bees thriving!

Chapter 4
Bee-Friendly Backyard Setup

Keeping it Simple and Stress-Free

When we first started beekeeping, we felt pressure to create the perfect bee-friendly garden overnight. We imagined our bees needed endless rows of colorful flowers right outside their front door. Good news—they don't!

Bees are incredible flyers and explorers. Did you know bees easily travel up to five miles from their hive searching for food, water, and pollen? That means your bees already have access to flowers and resources far beyond your backyard. So, relax! You don't have to transform your yard overnight, or even at all.

Beekeeping is a journey—slow, steady, and enjoyable. It's about letting nature unfold naturally, not forcing perfection from day one. Your bees know exactly what they need. Your job is to gently support their journey.

Best Places to Put Your Hive

Choosing the right spot for your hive matters, but it doesn't need to cause stress. Here are some easy guidelines to follow:

Sun and Shade: Bees prefer a mix of sunshine and shade. Placing your hive where it gets morning sun and afternoon shade is perfect. It keeps bees warm in the morning and cool during

Georgia's hot afternoons. More sun helps them keep predators at bay.

Protection from Wind and Rain: Pick a spot sheltered from strong winds and driving rain. A few trees, a fence, or even a shed nearby can make a great barrier. This keeps your bees comfortable and safe year-round.

Easy Access (for You!): Place your hive somewhere you can reach comfortably. You'll thank yourself later when doing hive inspections or harvesting honey. Remember our lesson—bees and honey are heavier than you think!

Safe Distance: Keep your hive a safe distance from high-traffic areas. Bees naturally fly in and out frequently, so a quiet corner of your yard works best.

Creating a Safe, Bee-Friendly Habitat (Without Stress!)

While bees will happily roam far and wide, planting a few pollinator-friendly plants in your yard can still bring joy and color close to home. But remember, you don't have to become a gardening expert overnight. Here's how to keep it simple:

- **Start Small:** Plant one or two flowering shrubs or perennials like lavender, salvia, or sunflowers. Each year, add a little more if you want—but only if you enjoy gardening!
- **Go Native:** Native Georgia plants naturally attract bees and other pollinators. Examples include milkweed, coneflowers, black-eyed Susans, and wild indigo. These plants thrive without much fuss.
- **Limit Chemicals:** Keep pesticides and herbicides out of your garden. Chemicals harm bees, pollinators, and beneficial insects. A chemical-free garden creates a safe haven for bees and your family.
- **Water Source:** Bees love shallow, gentle water sources. A small birdbath or a shallow pan with pebbles for bees to land on is perfect. Keep the water clean and fresh.

Remember, the key is low stress and high enjoyment. Create a simple, peaceful backyard you love—and your bees will love it too.

Bees Have a Mind of Their Own (And That's OK!)

One of the biggest lessons we've learned in beekeeping is that bees are wild creatures. They cannot be fully tamed or controlled—and honestly, that's part of their beauty.

You can guide bees, support them, and gently care for them, but ultimately, bees do exactly what bees want to do. They know better than we do what's best for their colony.

Sometimes bees swarm and start new hives, choose unexpected flowers over your carefully planted garden, or decide they prefer a different spot than the one you chose. And that's okay! As a beekeeper, it's important to accept that nature will always have its own plan.

The secret to joyful beekeeping is learning to trust and respect your bees, letting go of the need to control everything. Your job is to keep them safe, healthy, and happy—not to tame them. When you embrace bees' independence, beekeeping becomes a lot more relaxing—and rewarding.

Slow and Steady is the Way to Bee

Stress-free beekeeping isn't about getting it perfect on the first try. It's about taking small, manageable steps and enjoying each part of the journey. You don't have to rush to create a bee paradise in your yard. Your bees will thrive anyway, flying happily far beyond your property line to find exactly what they need.

Think of beekeeping as a slow and steady process. Each year you'll learn more, plant more (if you like), and understand your bees a little better. Eventually, without feeling rushed or stressed, you'll look around and realize you've created something wonderful.

Let Nature Be Your Teacher

Every season with bees teaches us something new. Some years

are full of honey, others are more challenging—but every year is worthwhile. Let nature guide you, and your bees teach you patience, joy, and acceptance.

As a beekeeper, your backyard becomes a place of quiet wonder, buzzing with life. Remember:

- Bees explore miles away—no pressure to plant the perfect garden!
- Start small, keep it chemical-free, and have fun.
- Bees are wild and wonderful—respect their independence.

Your journey with bees is about harmony with nature. Sit back, watch them fly, and enjoy this beautiful adventure.

Your bees already know how to bee—trust them to guide you as you learn.

Let's keep buzzing on to learn how you can keep your bees naturally healthy and happy in the next chapter!

Chapter 5
Keeping Bees Healthy (Naturally!)

My Chemical-Free Journey

When my husband and I first began taking beekeeping classes, we noticed something that bothered us right away—the heavy emphasis on chemical treatments. Our instructors repeatedly explained how chemicals were "necessary" to control mites and other pests. But something about this approach didn't sit right with me.

If these chemicals were strong enough to kill bugs (and sometimes even bees), I couldn't help but wonder, how safe was our honey? The common belief was that capped brood (bee larvae) and honey were unaffected by chemical treatments. But this didn't fully convince me.

Would you willingly eat something treated with chemicals strong enough to kill insects? We wouldn't either.

This discomfort led us toward a different path—natural beekeeping. I realized that everything was perfectly designed by God, and bees were no exception. Bees have survived thousands of years without humans putting chemicals into their hives, and I felt certain they could continue to thrive naturally.

Let's explore why chemical-free beekeeping makes sense, and how you can keep your bees healthy and strong without harmful treatments.

Why Go Chemical-Free?

Beekeeping without chemicals isn't just a trendy idea—it's a healthier way to raise bees and produce honey you can trust. Here's why chemical-free matters:

- **Cleaner Honey:** When you avoid chemicals, your honey remains pure, just as nature intended. You never have to question if your honey is safe for your family and friends.
- **Healthier Bees:** Bees treated with chemicals often become dependent on treatments, making their colonies weaker over time. Allowing bees to naturally defend their own hive creates stronger, more resilient colonies.
- **Balanced Environment:** Chemicals don't discriminate —they kill both harmful pests and beneficial insects, disrupting your hive's delicate ecosystem. Natural beekeeping supports balance and biodiversity within your hive.
- **Peace of Mind:** You can sleep better knowing your bees and your family aren't exposed to unnecessary toxins.

You see, God created bees perfectly capable of defending and managing their colonies. Honeybees naturally groom themselves and each other, removing pests and maintaining hive health. Given the right environment, bees use their instincts and teamwork to keep their homes healthy, clean, and thriving.

As a beekeeper, my goal is not to interfere excessively, but instead to gently support my bees. The less I intervene with unnatural substances, the stronger and smarter my bees become. It's about trusting nature, and understanding that bees were created with amazing abilities.

Simple Natural Ways to Keep Bees Healthy

Going chemical-free doesn't mean ignoring your bees—it means actively supporting their natural defenses. Here's how you can help your hive thrive without using chemicals:

1. Keep a Strong Queen

A healthy queen is the cornerstone of a healthy hive. Regularly checking your queen ensures she's laying plenty of eggs, creating strong generations of bees. Strong colonies are naturally resistant to pests and diseases.

2. Allow Bees to Build Natural Comb

When bees build their own wax comb (instead of using pre-made foundations), they naturally create cells sized perfectly to their needs. This reduces the likelihood of pests, like mites, taking over your hive.

3. Ensure Proper Ventilation

A well-ventilated hive prevents moisture buildup, keeping mold, fungus, and disease at bay. Good airflow helps bees regulate hive temperature naturally and keeps them healthier.

4. Choose the Right Hive Location

A sunny, dry location away from dampness and strong winds helps bees stay healthy. Proper placement reduces stress and protects the colony from potential threats.

5. Regular Inspections (but Don't Overdo It!)

Regular inspections help you identify problems early without using chemicals. However, too many disruptions can stress your bees. Balance careful observation with respectful distance.

Natural Pest Solutions (That Actually Work)

Going chemical-free doesn't mean leaving bees defenseless. Here are effective natural solutions we've successfully used:

- **Powdered Sugar Dusting:** Gently dusting your bees with powdered sugar encourages them to groom each other, naturally removing mites. It's safe, easy, and non-toxic.

- **Essential Oils (Thyme, Lemongrass, Peppermint):** Small amounts of essential oils can safely repel pests without harming bees or contaminating honey. Always use sparingly and carefully, following trusted guidelines. 31
- **Screened Bottom Boards:** These special hive bottoms allow mites and debris to fall through, keeping your hive cleaner naturally. It's a simple yet effective improvement to your hive setup.
- **Break the Brood Cycle:** Temporarily removing frames of brood disrupts pests' breeding cycles. It's a gentle, natural way to reduce mite populations without chemicals.

Embracing Natural Beekeeping

When we first questioned chemical use, some experienced beekeepers were skeptical. But our experience confirmed something profound: Bees thrive when allowed to do exactly what they were created to do. Nature, guided by God's wisdom, always knows best.

Natural beekeeping isn't complicated. It simply requires a change in perspective. Instead of trying to control every aspect of your bees' lives, let nature guide your decisions. You'll discover stronger bees, cleaner honey, and a much more rewarding experience.

Your Bees, Your Choice

I encourage you to think carefully about chemicals in your hive. If something doesn't feel right, trust your instincts—just like we did. The reward is pure honey, a stronger colony, and the satisfaction of knowing you're supporting nature's perfect design.

You started beekeeping to connect with nature and support the environment. Chemical-free practices help you achieve exactly that. Let your bees teach you how wonderfully designed they are.

It's time to trust your bees to do what they do best.

Chapter6
Harvesting Honey
(Easy and Enjoyable)

The Sweetest Reward

Harvesting honey is probably the most exciting part of being a beekeeper. Nothing beats tasting your first spoonful of pure, golden honey straight from your hive.

It's a delicious reminder of all your bees' hard work and a joyful celebration of your partnership with nature.

But when is the best time to harvest honey here in Georgia, and how do you know your bees are ready? Let's explore these simple signs from your bees and nature that tell you it's harvest time.

Georgia's Honey Harvest Seasons

In Georgia, you'll typically find two main honey seasons:

Spring Honey Flow (Late April to Early June): This is the primary honey season when everything is bursting into bloom. Flowers, shrubs, and fruit trees like peaches, blueberries, blackberries, clover, and wildflowers provide a bounty of nectar. During this period, bees work extra hard, filling your hive quickly. Most beekeepers harvest honey around mid-to-late June when the honeycombs become fully capped, signaling the bees have ripened the honey perfectly.

Late Summer/Fall Flow (August to September): Some Georgia beekeepers experience a second smaller harvest, often called the "goldenrod" or "wildflower" honey flow. During late summer, plants like goldenrod, asters, and various wildflowers bloom, giving bees one more nectar boost before winter. If the season is strong, you might harvest honey again around mid-to-late September.

Signs It's Time to Harvest Honey

Your bees and the environment will clearly show you when it's time to harvest. No guesswork needed!

1. **Bees Cap Their Honeycomb:** The clearest sign is seeing capped honeycomb frames. Bees cover (or "cap") honeycombs with a thin layer of white beeswax when honey is fully ripened. This means moisture levels are perfect, making honey ready for harvest. Aim to harvest frames that are at least 80%capped to ensure perfect, stable honey.
2. **Hive Activity Increases:** During peak nectar flow, you'll notice your bees seem busier than ever. Your hive will hum with activity as bees race in and out, bringing nectar from blooming plants. When hive traffic slows down and the bees start capping combs, it's time to think about harvesting.
3. **Plants and Flowers Signal Honey Flow:** Observing your local environment gives you important clues. Look for blooms of flowers like clover, honeysuckle, wildflowers, and blooming trees such as tulip poplar, magnolia, and black locust. When these blooms fade, it's usually a good indicator your bees have stocked up plenty of honey.

Easy Steps for Harvesting Honey

Harvesting honey can seem intimidating your first time, but it's actually straightforward—and lots of fun. Here's a simplified approach:

Step 1: Choose a Sunny, Calm Day

Pick a clear, sunny day for harvesting, preferably midday. Most bees will be busy foraging, leaving fewer bees in the hive, making the process calmer and easier.

Step 2: Gently Smoke the Hive

Use your smoker lightly to calm the bees. This encourages them to move deeper into the hive temporarily while you remove honey frames without stress or disruption.

Step 3: Carefully Remove Frames

Gently lift out fully capped honey frames one at a time. Give each frame a gentle shake or brush to remove bees. Take your time—there's no rush.

Step 4: Uncap and Extract

Back indoors, carefully slice off wax caps using a simple uncapping knife or fork. Place your frames in a honey extractor (or crush and strain if you prefer a simple method). Spin or drain your honey gently into clean containers.

Step 5: Let Honey Settle

Allow your freshly harvested honey to sit a few hours or overnight to settle naturally. Bubbles and wax bits rise to the top, making your honey clearer and smoother when bottled.

Step 6: Bottle and Enjoy!

Finally, pour your honey into clean jars, label them proudly, and enjoy sharing with friends, family, or your community.

Keep it Pure and Simple

One of the joys of beekeeping naturally is knowing exactly what's in your honey—nothing but pure nectar, lovingly made by your bees. No additives, no chemicals, just pure Georgia goodness.

Remember to keep harvesting stress-free. Harvest gently, calmly, and joyfully. The less stress you put on your bees (and yourself!), the happier everyone will be.

Sharing Your Honey: Family, Friends, and Community

Honey harvest time is also a perfect opportunity to share the joy of beekeeping with others. You'll soon discover nothing brings people together like sharing jars of your own delicious, pure honey. It makes wonderful gifts and is a great conversation starter to encourage others toward chemical-free practices and bee conservation.

Each jar is a sweet reminder of your commitment to bees, your community, and the environment. Use your honey harvest to inspire others and spread the message that natural beekeeping truly works—and tastes amazing too!

Harvesting is About Partnership and Gratitude

Harvesting honey is more than simply collecting a sweet reward—it's a beautiful reminder of your partnership with bees and nature. Your bees have worked tirelessly, pollinating plants, enriching our food supply, and generously providing extra honey.

Be thankful for your bees' hard work and respect their efforts by harvesting responsibly. Always leave enough honey in the hive for your bees to feed themselves through winter—usually 30-50 pounds in Georgia is a good rule of thumb. Your bees deserve it!

Enjoy your harvest. Let it be a joyful celebration of patience, simplicity, and gratitude. Your bees have given you the sweetest gift—share it widely, smile proudly, and keep Georgia buzzing!

Now that you've harvested honey, let's explore how you can become a passionate advocate for bees, encouraging others to protect pollinators and create healthier communities together

Chapter7
Becoming a Bee Advocate

Bee the Change You Want to See!

Congratulations—you're well on your way to becoming an amazing beekeeper! But your journey doesn't have to end here. By now, you've discovered how incredible bees are and the vital role they play in our food systems and communities. Now, let's take that passion and spread it around!

Advocating for bees doesn't have to mean big, complicated projects or grand speeches. Even the smallest actions make a powerful difference. Trust me—every little thing you do matters.

Why Advocate for Bees?

Bees need your voice because they can't speak for themselves. Right now, bees are facing challenges like habitat loss, pesticide exposure, and declining flower sources. The good news? Simple actions by people like you and me can turn things around.

Imagine if every backyard in Georgia had just one hive, or even just one small patch of flowers. Together, we could build healthier neighborhoods, feed families with pure honey, and create stronger communities. That's real power, and it begins with you.

Supporting Pollinator Habitats (Every Yard Counts!)

Whether you live in the country or a busy city, creating pollinator habitats is easier than you think. Remember, you don't need acres of land—just a few simple actions can transform your yard, balcony, or community garden into a bee-friendly haven.

Easy Ways to Help:

- **Plant a Flower (or Two!):** Even planting a single pollinator-friendly flower in your yard or a pot on your porch helps. Bees don't discriminate—every bloom counts!
- **Avoid Chemicals:** Choose natural ways to manage weeds and pests. Every pesticide-free yard is a safe refuge for bees.
- **Provide Fresh Water:** A small dish of water with pebbles gives bees a safe spot to drink, especially in hot Georgia summers.
- **Let Grass Grow (a Little):** Letting a small patch of your lawn grow naturally supports bees, butterflies, and other beneficial insects.

Imagine if every family did just one thing from this list. We could transform communities overnight!

Promoting Chemical-Free Practices (Start Small, Dream Big)

Chemical-free beekeeping changed my life, and it can change yours too. By sharing your story, even casually with friends or neighbors, you can inspire others to choose safer, healthier practices too.

Here's how:

- **Talk About Your Honey:** When sharing honey, talk about how you raised bees naturally. People love stories, and yours is powerful!
- **Encourage Neighbors:** Suggest easy alternatives to harmful chemicals, like natural pest management. Most people are open to safer methods once they know about them.
- **Support Local Organic Farmers:** Buying and recommending organic food strengthens local farms and reduces chemical use in agriculture.

Small conversations matter because they plant seeds. You might inspire your entire community toward healthier, chemical-free lifestyles!

Simple Steps Anyone Can Take

Advocacy doesn't have to be complicated. Here are easy ways to make an immediate impact:

- **Share Your Journey on Social Media:** Snap pictures of your hive, your flowers, or your delicious honey. Tag friends and share your natural beekeeping experience. You never know who you'll inspire!
- **Educate (Gently!):** Leave helpful brochures or a jar of honey at your local coffee shop, library, or farmers' market with a friendly note explaining why bees matter. Simple acts spark big conversations.
- **Donate Honey to Good Causes:** A jar of honey at a community fundraiser or charity event spreads the bee-love. Include a small card explaining your chemical-free approach.

Bees Bring Communities Together

One beautiful thing about bees is that they naturally create community. Just one hive in your backyard can inspire neighbors to plant more flowers, use fewer chemicals, and become more mindful of nature.

Bees show us that even the smallest creatures have a mighty impact when they work together. Let's follow their example, building community around our shared love for nature, healthy food, and a better world.

Every Action Matters—Especially Yours

Advocating for bees is about believing that small steps can create big changes. Every flower planted, every chemical avoided, every conversation had—all these small actions ripple outward, creating a healthier environment for bees and people alike.

Imagine your community filled with buzzing bees, blooming flowers, and healthy food. This is possible. And it starts with you—today, right where you are.

So let's keep it simple, positive, and joyful. You've come so far as a beekeeper, and your next step—advocacy—is easy. Share your love for bees, tell your story, and never underestimate the power of your voice.

Together, let's keep bees buzzing. Georgia depends on it, our families depend on it—and our future does too.

Your bees thank you, your community thanks you, and future generations will too.

Now go bee amazing!

Conclusion: Your Journey Has Just Begun

Celebrating Your New Adventure

You've made it! You've traveled through this beginner-friendly guide, and now you're ready to start your own beautiful adventure into backyard beekeeping. But here's a secret—your journey is only just beginning.

Beekeeping is not something you master overnight. It's a journey of discovery, connection, and continuous learning. Each season offers new insights, fresh lessons, and unexpected joys. As you step forward, always remember this: beekeeping is more than just a hobby; it's a powerful way to reconnect with nature, support your community, and live a healthier, more mindful life.

From Beginner to Advocate

When my husband and I started our beekeeping journey, we had no idea how deeply it would impact our lives. As disabled veterans, we quickly realized that keeping bees meant far more than simply harvesting honey. It became a pathway to health, a source of joy, and a meaningful way to give back to our community.

We discovered how powerful simple, chemical-free practices could be. We learned firsthand that nature, guided by GOD's perfect design holds all the solutions we need. Our journey taught us that simplicity is key, patience is essential, and the best rewards often come slowly and quietly.

Now, you're beginning your own version of this incredible story. Like us, you'll discover lessons about bees—and life—that books alone can't teach. Your bees will guide you, your community will inspire you, and your love for beekeeping will grow deeper every year.

Getting Back to Basics (What's Next?)

This guide has been about getting you started—without stress or complexity. But there's so much more to explore. In fact, our next journey together will take us deeper into the beauty of getting "back to basics."

We'll dive into simpler methods—not just for beekeeping but for other areas of agriculture and gardening. We'll talk about returning to traditional ways of raising healthy food, tending plants, and caring for animals naturally. Just as bees have shown us the beauty of simplicity, we'll apply these lessons across all areas of sustainable, chemical-free living.

You see, beekeeping opens a doorway to a simpler, more intentional lifestyle. It teaches you to slow down, observe carefully, and respect nature's perfect systems. This mindset will enrich your life far beyond your backyard hive. Together, we'll keep learning, growing, and discovering that nature truly does know best.

Building Strong Communities, One Hive at a Time

One of the greatest joys of beekeeping is the community it creates. As you start your journey, don't hesitate to share your experiences with others. Talk about your hive, offer tastes of your honey, and encourage your neighbors to consider chemical-free gardening.

Remember, every hive in Georgia contributes to healthier local agriculture. Each hive means more fruits, vegetables, and flowers. It means more pure honey, less chemical exposure, and stronger neighborhoods. Your bees are quietly improving your community every day, and you're a vital part of that success.

Imagine if every neighborhood had just a handful of backyard beekeepers. Imagine local farms thriving without harmful pesticides. Imagine communities where families enjoy pure, healthy honey right from their neighbor's backyard. This future is achievable—and it begins with you.

Trusting Nature and Trusting Yourself

Throughout this guide, one message shines through clearly: trust nature and trust yourself. Your bees are perfectly equipped to thrive when given simple, natural support. Trust them to guide you as you learn. Trust yourself to make decisions that align with nature's wisdom.

When you feel unsure, simply pause and observe. Nature provides clear answers if you're patient enough to listen. As your bees have shown you, great strength comes from gentle, steady care—not from rushing or forcing outcomes.

The Gift of Gratitude

As your first season unfolds, you'll experience countless moments of awe and gratitude. Harvesting honey, watching bees dance across your garden, tasting fresh honeycomb—these small joys will enrich your life deeply. Remember to savor each moment and always thank your bees for their incredible work.

Express gratitude not just in words, but by your actions. Continue advocating for pollinator habitats, chemical-free gardens, and simple, sustainable practices. Each grateful action multiplies your positive impact, transforming lives beyond your own.

You Are Ready

You might feel nervous or unsure about stepping fully into beekeeping. That's okay! Every great beekeeper started exactly where you are today—with curiosity, enthusiasm, and perhaps a little uncertainty. The wonderful news? Bees are excellent teachers. Follow their lead, remain open-hearted, and enjoy every discovery along the way.

A Promise to Keep Growing Together

As your beginner guide concludes, our shared journey is just starting. I promise to keep learning alongside you. In future books, we'll dig deeper into the beauty of simplicity, the strength of nature, and the importance of returning to basics across agriculture and life.

Let's keep connecting, learning, and growing together. Let's encourage each other, support our communities, and become examples of healthy, sustainable living. Bees started our journey— but together, we'll explore even further.

You've Got This!

You've chosen to become a beekeeper. That's no small thing. You've chosen to care for bees, protect your environment, and reconnect with nature. You've chosen a healthier, simpler path that enriches your community.

So, take a deep breath, smile, and step forward confidently. You're ready. Your bees are ready. Your community needs you.

Your beekeeping journey has truly just begun, and the future is buzzing with possibilities. Let's keep bees thriving, communities healthy, and Georgia beautiful.

Thank you for choosing this path. I'm proud of you, grateful for you, and excited to see the wonderful things you'll accomplish.

Keep buzzing—your bees, your community, and your future depend on it.

I'll see you soon, friend, in our next chapter together— getting back to basics and growing something beautiful, naturally.

Our Layens and Langstroth hives can be visited at the G.W. Carver Victory Farm and Garden in Columbus, GA

Layens Hives

Langstroth hives

Resources and Further Reading

Congratulations—you've reached the end of this beginner's guide, but your learning doesn't have to stop here! Below are some excellent resources to support you as you continue your backyard beekeeping journey. Each resource has been chosen to help you dive deeper into chemical-free practices, horizontal hive management, and community-driven beekeeping.

Recommended Books:

- **Beekeeping with a Smile** by Fedor Lazutin *A fantastic, easy-to-read book that inspired our natural, horizontal hive approach. Lazutin shares practical wisdom about simple, chemical-free, low-maintenance beekeeping.*
- **Keeping Bees with a Smile: Principles and Practice of Natural Beekeeping** by Leo Sharashkin *Written specifically for backyard beekeepers using horizontal hives. Dr. Leo explains clearly how simplicity and natural methods lead to happier bees and beekeepers.*
- **The Backyard Beekeeper** by Kim Flottum *A popular beginner's guide filled with great illustrations, easy explanations, and practical tips for first-time beekeepers.*

Online Resources:

- **Foundation Honey Co YouTube Channel** https://youtube.com/@foundationhoneyco *A valuable resource created by experienced Georgia-based beekeepers (us!), highlighting natural beekeeping methods, horizontal hive management, and community-focused bee advocacy. Perfect for visual learners!*
- **Georgia Beekeppers Association** https://gabeekeeping.com *Local resources classes, mentorship opportunities, and ways to connect with other Georgia*

beekeepers.

- **Honey Bee Suite** https://honeybeesuite.com *A clear, engaging, and informative blog covering all aspects of natural beekeeping, from managing pests without chemicals to understanding bee behavior.*
- **Xerces Society for Invertebrate Conservation** https://xerces.org *A great organization focused on protecting pollinators through chemical-free gardening, creating habitats, and community advocacy.*

Local and Community Groups:

- **Find your Local Beekeeping Club** *Google "local beekeeping clubs near me," or check with your local Extension Office to find supportive groups and workshops in your community.*
- **Farmers' Markets and Community Gardens** *Visit local markets and gardens to connect with like-minded folks who care about bees, pollinators, and chemical-free agriculture.*

About the Author

Radiah Mallard is a passionate beekeeper, veteran, and community advocate living in Columbus, Georgia. Alongside her husband—both proud disabled veterans—Radiah began her beekeeping journey seeking simplicity, health, and connection to nature. Her combat experience and jumping out of airplanes and her husband's combat injury taught them resilience and adaptability— qualities that served them well as they transitioned from traditional Langstroth hives to the ease and simplicity of horizontal (Layens) hives.

Radiah and her husband founded **Foundation Honey Co.**, a nonprofit dedicated to promoting natural, chemical-free beekeeping practices while supporting underserved communities. They are committed to teaching others about the critical role bees play in food security and healthy ecosystems, particularly in communities affected by food deserts. They also founded **D'Borah Concepts LLC** , a veteran-owned company that creates faith-based books to inspire spiritual growth and introduce urban communities to the enriching world of agriculture.

In addition to beekeeping, Radiah finds joy in relaxation through spa treatments, nature walks, meditating on the Word of YAHUAH, and encouraging others to embrace a peaceful, Christ-centered lifestyle. She deeply believes in trusting in GOD, honoring creation's perfect design, and advocating passionately for sustainable, chemical-free agriculture...amongst other things!

To learn more about natural beekeeping and horizontal hives, visit Foundation Honey Co.'s YouTube channel at youtube.com/@foundationhoneyco.

For more books or questions about beekeeping, visit https://stan.store/DborahConcepts

To request speaking engagements, visit https://dborahconcepts.co/radiahmallard.

Together, let's keep bees buzzing and communities thriving!

NOTES

Made in the USA
Monee, IL
13 April 2026

48266812R00030